The Mobility Backhaul Report

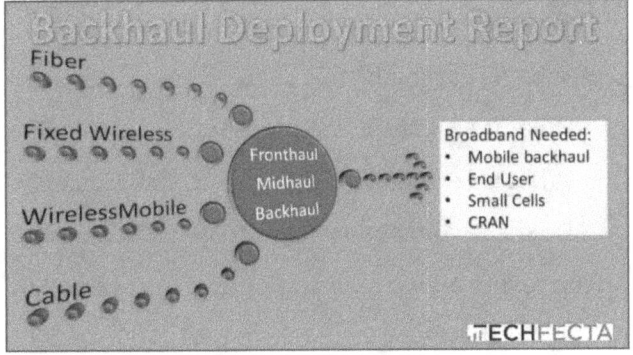

The Mobility Backhaul Report

Contents

The Mobility Backhaul Report 1
Copyright 5
Thank you! 6
Introduction: 6
The Overhaul of all the Hauls! 8
    Overview: 8
    What is Backhaul? 13
    What is Midhaul? 18
    What is Fronthaul? 19
    Do these connections have to be Fiber? 20
Fiber overview: 22
    Installation: 22
    Permitting: 24
    Competition: 25
    Reoccurring fees: 25
    Who benefits from the fiber growth? 26
Wireless overview: 28
    Spectrum: 28
        Spectrum for 5G Fixed Wireless 32

- System distinctions: 37
- Costs in wireless: 38
- Resources: 39
- What is Fixed Wireless? 42
  - Fixed Wireless Overview 47
  - Who will provide broadband? 58
- Why does Fixed Wireless Matter for City Growth? 60
  - Fixed Wireless could be a key to Growth for Urban and Sub Urban areas. 62
- Unseen Costs of Deploying Fiber and Wireless 68
  - The rise of Site Acquisition...........................68
  - Wireless Costs 71
  - Before you mount anything: 73
  - Drawbacks of Running Fiber 73
  - No matter what you deploy: 75
  - Rent Applies to Everything...........................77
  - Summary of costs: 78
- Broadband Initiatives for Cities........................80
  - Primary 85
  - Partial 87
  - Facilitator 90

- Not all Deployments are a Success!............92
- Smart City Investment 94
- Who buys broadband, really?..........................97
  - Pricing matters!..99
  - When Incumbents fight Back!....................100
- What about Wi-Fi?..103
  - Some States Prohibit Public Networks! 105
  - City Strategies for a Broadband Initiative..107
- What will the future hold?..............................109
- Acronyms and Definitions..............................113
- Thank you! 116
- Copyright...116
- About Wade 117
- Learn more!...118

Copyright

First Edition © 2018 by Wade Sarver. All rights reserved. No part of this publication may be reproduced, stored in a retrieval system, or transmitted in any form or by any means, electronic, mechanical, photocopying, recording, scanning, or otherwise, except as permitted under Sections 107 or 108 of the 1976 United States Copyright Act, without the prior written permission of the author.

I am not a lawyer or an actively certified safety expert. This book is completed based on research and my experiences. Safety processes and procedures are constantly updated and improved over time. The material contained is for reference only and may include products, information, or services by third parties. I do not assume responsibility for any third-party material referenced in this book.

This document is a guide to help people and not a guarantee that you will do everything properly. By reading this, you agree that myself and my company is not responsible for the success or failure of your business decisions relating to the information presented in this guide.

www.wade4wireless.com

www.techfecta.com

Cover and design by Wade Sarver

Thank you!

Thank you for purchasing this report, I appreciate your support. I pray that it serves you well.

If you can, let me know what you like and didn't like about this report. What should I add, and what should I skip next time.

If you need one on one consulting or specific reports, feel free to reach out at wade@techfecta.com for direct support.

Introduction:

This Mobility Backhaul report covers the connection between all the equipment in the field. Specifically, small cells and macro sites connecting to the core and out to remote radio heads.

To make things easy for you, I added Acronyms and Definitions to the end of this article. That may help if you get confused or want clarification.

Notice that I have made a resources section to help you see where I get my information and how I came to the conclusions I have. This way, if you want to, you can research your specific topic with greater depth.

The Overhaul of all the Hauls!

Overview:

While this report stresses backhaul because technically, it's all backhaul. There are many aspects beyond backhaul. We have fronthaul and midhaul. I thought we should have an overhaul of the naming convention. It helps to understand what each one is.

One of the significant barriers to rolling out wireless sites has been the backhaul. You would think that fiber is everywhere, but when it comes to connecting to fiber, it can run into some money. When deploying with fiber and attaching to a pole or remote location, it's not cheap. (Small cells are commonly put on poles along the street.) If there is not existing fiber nearby, it cost a lot of money. If the fiber at a location is maxed out, then it costs a lot of money. See the pattern?

Also, if you need a lot of bandwidth, fiber costs a lot of money each month. So, the CapEx of the installation could be expensive as well as the OpEx of usage every month.

Don't get me wrong; fiber is incredible! We all love the fact it can handle so much bandwidth in a pair of fiber strands. Amazing! It put a dent in the wireless backhaul market because it is rolling out everywhere and quite flexible. We all love fiber. However, it's not cheap to install or to pay the monthly reoccurring.

Remember when wireless backhaul was great? We loved wireless backhaul back in the day because we could pay $10K to $20K to get it running and it would be there for years. It took up tower space, but it was reliable and a "pay-once" type of deal. Sure, you had maintenance, but no monthly charges other than tower rent. It was a fine solution for public safety and cell phones. It worked well for a long time and if you designed the system right, you had redundancy.

Oh, but then we had a bandwidth explosion. Well, it was hard to add bandwidth to existing microwave setups, if you could at all. Modern networks demanded more than the long-range wireless could supply. It's too much for them to handle. So, now we get fiber, and we may use wireless as a backup, but the traditional 6GHz and 11GHz links can't provide the Gbps links we need today. The broadband requirements are growing, so the idea of putting in wireless links seems to limit growth. More on that below.

What can we do? Well, the release of the fixed wireless spectrum may solve this problem. If this is something that can grow along with the needs of the end user, then it is going to be the Midhaul solution. This would be the link between a fixed radio head and the controller or core. See the illustration below. We need to look at the fixed wireless as the midhaul and the fronthaul. We also need to look at fiber as more than the backhaul solution. It could be the link for the edge to get to the internet or the midhaul or the fronthaul.

All these connections need to be made. As we add hops, we also add latency. Think of how the small cell or remote radio head could connect to the core and to the internet simultaneously. There may be more than one link at a site.

If the small cell or remote radio head needs a direct connection to the internet, it may not need to be a fiber link. It could be just something to offer low latency, so any type of internet connection may be just as good. The idea of that connection is to lower latency, so bandwidth may not be the issue. So, order accordingly, remember that we need to be cost conscious when planning.

Backhaul is the connection to the internet or the core. The core is the hub where all the mobile equipment lies.

Midhaul could be the link between the controller or the radio head that feeds the next link.

Fronthaul is generally the link between the controller and the radio head or small cell. It could be the link from the radio head to the UE device. Fronthaul should be the final link, but not the last 200 meters.

All the same, we look at the backhaul using all means necessary to make the connection to the final radio. It could be a combination of several links that act as a chain to get the data from the end user to the core and eventually its destination. Each network will be responsible for moving data from point A to point B using any means necessary. It all works together to ensure that the end user gets what they ask for.

Cost-effective solutions are what we want. It is not always fiber. It would be any combination of wireless and fiber. As long as it is reliable and allows for growth. Growth is critical in today's world of expansion.

Being cost-effective means that we need a balance between the payback, (number of subscribers) and the spend, (installation and monthly costs). That is only part of it. We need to know, what is available? If the fiber is not available, you may need wireless to get the site on the air. If wireless is not available, you may need to move the site to another location where something is available. Most times moving across the street can make all the difference. Availability is vital because if you need to run fiber across a street, it may involve trenching and permitting, a hefty cost for installation.

Reliability is a crucial factor here. In the past, wireless would show errors during a rainstorm. This was a problem because the link would have hi bit errors. The rain was a problem. Fiber could get knocked down if it's overhead, another issue that has caused problems in the past. Make sure your solution is reliable.

So, let's look at backhaul, midhaul, and fronthaul as one. After all, it's all the means to an end. They are all needed to get the data where it must go, both ways. No matter what the link, it is part of the solution. It takes planning. All I am asking is that you need to be open-minded. We often look at fiber as the only solution, but there are more than one means to this end. We have options, and they are growing every day. Let's take advantage of what we have and think outside the box. Fiber or wireless, it really doesn't matter if it fills the

Backhaul, Midhaul, and Fronthaul.

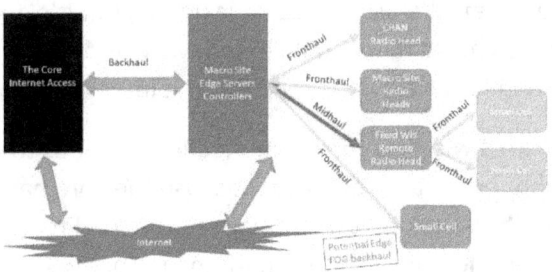

needs we ask for. As long as it meets the criteria to connect the end user to the core.

What is Backhaul?

Backhaul is the connection back to the source, like the internet or the core. Technically, everything I describe in here could be backhaul, but the purpose is to get the signal back to the core or data center. This is the key connection to the internet.

When you look up backhaul, it is the network that connects to the backbone. I look at it as more than that. We need to see backhaul for what it is, a connection to a controller somewhere as well as a connection to the internet.

It's ironic, but in mobility, we can't do anything without the backhaul. The thing is, it is the lifeline to any remote connection. Even your home needs backhaul to connect to the internet. We used to call it last mile, but it's much more than a mile in today's world. We need to have a connection to the internet.

With a wireless carrier connection, it needs to connect to a core. This is because the core has all the controllers and databases needed to keep the outside sites operating properly. While they can operate on their own, remote sites need the information that the core is providing. This is so that the proper customers get the connectivity they paid for and customers on other system don't. This keeps the billing in line and opens up internet access to these devices.

Home connections are a different story, all you need there is internet access. Chances are most people have a cable modem, maybe DSL. These connections go to a data center somewhere, normally. You may have Wi-Fi in your home making you think you have a wireless backhaul, but it connects to a wired backhaul somewhere. Even if you use your smartphone for internet access, it connects to a site somewhere that has fiber.

The end user will connect somewhere so the provider can track the user. In wireless, it's a little more complicated. The core must track all users on their wireless devices for connection and tracking purposes. They track them so that they know what kind of minutes or data they use as well as how to route a voice call.

The core of the wireless system, which generally is regional, will also have internet connections which means that even they have a backhaul. You have a backhaul for the backhaul. The core has a connection to the internet. Through that connection, it connects to all the sites that it controls, in some way.

Everything connects to the internet somewhere. At one time these were the data centers, the cores, the hubs where everyone had a direct connection to the internet backbone. Today it's different. Today we want to offload those pipes. Offloading the traffic has changed the way that wireless carriers do business.

The backhaul was becoming so overloaded that it was causing problems. The carriers wanted to track and control all the data at one time, but it became too much. They could not handle the traffic, especially with new data plans. Things had to change. Today they welcome offloading. The rollout of 5G and network slicing help that even more. With the growth of MEC, Mobile Edge Computing, it will offload even more traffic. So now the backhaul at a site not only connects to the core but has a route directly to the internet. However, it adds complexity to the network.

Who knew the internet would be so popular that nearly everyone would rely on it. Offloading became a key component in the new mobile backhaul network. It's not the only thing, with 5G we're required to lower latency and provide even more bandwidth to the users. All this while improving the user experience.

The mobility backhaul I talk about here is from the BBU or small cell to the core. However, today we have broken it up even more. The backhaul could be to the core or maybe just an internet connection. It could be a VPN connection through an internet connection.

All the same, the physical backhaul is generally fiber. If we use wireless, it's only a partial solution to the over backhaul. The backhaul is generally an entire network. In most cases not just a straight shot to the core, but a series of connections to a regional core. To be clear, most of those connections ride across fiber. Fiber is a key component to growing backhaul and internet access. If you work backhaul, then you love fiber.

Fiber allows growth. There was a time when we thought a pair of fiber would be all we ever need. How naive I was to think that! It takes more and more fiber to handle the bandwidth required to serve customers' needs today. Think about what we will need for the future. We must have more fiber pairs to grow effectively. That means the fiber we have run now will to be sufficient. Carriers want their own backhaul as a site. The shared backhaul systems that were introduced years ago did not do well. Every major carrier wants their own backhaul without anyone else on it. They do not want to share, and they do not want to play with others unless they get paid for it. Basically, they want control.

Fiber allows this. When installing the backhaul at the site, it pays to have plenty of dark fiber to allow for growth. It's critical to plan. I have seen sites in Manhattan, NYC, that did not run enough fiber and it severely limited growth at the poles throughout the city. More than one carrier wanted on a pole. The city tried to limit the carriers but found out that they need to share and be as effective as possible. The alternative was to run fiber to every pole, which they may do someday soon. Carriers do not want to share fiber runs or even fiber cables in most cases. That means we need even more fiber run to each pole or site.

When looking at how much will be needed, think of this, massive MIMO sites use up to 3 times the fiber that a traditional site does. Sites that had a 1Gbps backhaul will need 3 to 5 times as much. It is not going to stop anytime soon. The demand for broadband is still growing. How much is enough? We don't know yet. All we know is that we need more and more.

The key is to plan for growth and then realize you need more.

What is Midhaul?

This is tricky and new. The Midhaul is a new concept, and it makes sense. What is happening is that the OEMs are trying to help the carriers understand that the connection may be put together in hops, not as one huge backhaul. A Midhaul might allow the cloud connection somewhere between the core and the last small cell or radio head. It may allow a wireless link to make a connection where there is no fiber. It may be where the backhaul feeds several sites, and the Midhaul could connect 2 to 3 more sites. It is a component that we didn't break out before these systems relied on BBU hotels, small cells, CRAN, and specifically FOG computers in the mix.

The 5G wave will be the thing that makes EDGE and FOG computing a necessity in these networks. They will be changing the way we look at the wireless system. It's not all about the backhaul and radios anymore. Now we need more intelligence at the wireless site along with more connections back to the internet in some way. This is how we do it, we put a server out there for the EDGE or FOG, and we give it a dedicated connection to the internet to lower latency. To make things clean, it can route things out of the existing system direct to where it must go.

This breaks up the backhaul and fronthaul and allows a new leg to be entered into the system.

What is Fronthaul?

Here we have the last mile connection to the wireless site. We call it fronthaul because it generally is the connection from the BBU or controller to the remote radio. The remote radio generally is just a radio head, as in the case of CRAN or cRAN, but it could be a small cell. All we need to know is that the fronthaul now could be more than a mile as the systems begin to mature. The thing is, most permits see the end connection as a small cell, but it could be a higher power radio head located as part of the cloud RAN or Centralized RAN.

The fronthaul generally needs to be as low latency as possible. Especially if the carrier is using it. However, small cells allow for more latency. It all depends on the application.

There are wireless systems for the fronthaul, but in most cases, fiber works best if there is a CRAN.

Keep in mind that a small cell is a full cell site, small, but in one unit whilst the CRAN or cRAN is relying on a remote BBU to control it. The CRAN needs to be talking to the BBU as close to real-time as possible so that it can get every command. Latency is crucial to the CRAN system. That is why fiber is the preferred choice for CRAN.

Do these connections have to be Fiber?

No, of course not. However, fiber is the first choice for backhaul because of the bandwidth and expansion available in fiber. Most times fiber is going to be used for backhaul because of its versatility and growth potential.

However, wireless was a viable option at one time. The problem with the older radios is that they could not grow or add bandwidth in the spectrum they were allocated. This is still an issue with many radios today. Most times you have to add a new radio.

One more thing, fiber is needed somewhere. Wireless can take you so far. Eventually, you need a connection to the internet and the core. It all depends how you want to mix and match the connections.

Fiber overview:

Fiber is needed somewhere in the network. We all rely on fiber at some point to take us home to the core and make the connection to the internet. It's not needed everywhere. Deployment engineers start to assume that they need fiber everywhere. That is where the problems begin. I admit that time has changed, and we seek fiber everywhere, but that's not always possible.

I think that Sprint makes progress when trying to use UE backhaul more and more. While this did not work out so well for other reasons, it did make progress in using UE backhaul in more and more small cell deployments which turned out to be cost-effective and quick to install.

While fiber generally is the first choice, it's not the only choice, in 2018 and 2019 it's not. This may change.

There are drawbacks to using fiber. To cover the obvious thing, Fiber can be very expensive to install. Why? Let me count the ways.

Installation:

- **Trenching** – if you must trench, fiber becomes very expensive, and not just for the labor to do it. The labor is not the most expensive part unless you're going under pavements or concrete. Even then they came up with boring techniques to go under many obstacles. The thing that causes delays in trenching is the permitting process. Almost all cities and towns have a process to file for permits, for a fee, and have the right to tell you now or they may have a dig once policy. More on that later.

- **Overhead** – you would think that mounting on an overhead pole would be quick and easy. It is from an installation standpoint. Again, installation is not why it's so expensive or takes a long time to get done. In fact, when you get to the installation phase, it moves along quickly. There are several things here that slow the process. Again, permitting must be done, and it takes time for approvals and to get the permits. However, the real issue with the poles is ownership of the pole. Who owns it and who will allow you to be on it. You see, many cable companies have been on these poles for years. The city or the utility company may own the pole, but an incumbent on the pole may have put in their lease that they can deny future competitors access to the poles. If you don't think this happens. It happened to Google Fiber repeatedly. It happened in Nashville, a city that so desperately wanted Google Fiber, but AT&T and Comcast did all that they could to keep Google off any of its poles, with lawsuits and lease restrictions. I have the link below in the resources area for this section. The local government thought they could overrule the agreements that AT&T and Comcast had on the poles, but the local ruling was overturned in US District Court. AT&T and Comcast banned Google from using their poles.

Permitting:

This is what costs money and time. While most companies can work with the local government, not all governments will help speed things up. All the same, it costs money to get through the process. I don't blame the local governments. It is a money maker for them, but they need to regulate how things are done so that it makes sense and looks aesthetically pleasing. The government needs to have control in some way; I just wish they could do it faster.

Competition:

This is what I mentioned in the installation section. If the competition has the rights to the poles and they can deny anyone to using them, what can you do? You find another way to get in. The incumbent has a lot of power with local governments. Comcast is a very powerful and rich company.

How can the competition hurt you? Let me count the ways:

- Ban you from using their local poles.
- Influence local government to make it harder for newcomers to deploy and permit.
- Lower their prices to drive the newcomers' costs down.
- Upgrade their system to force newcomers to come in strong.
- Lock up local contractors to raise costs for the newbies.

Reoccurring fees:

Here is what fiber providers love, that monthly income. They know that it costs money to do all the above, so they make it back in the monthly income. This is where the carriers must pay for the service. It's not new, look at any service you have, the chances are good that you pay for them monthly or annually.

Don't get me wrong, just because they don't like the service fee; they get a lot for it. The provider has to maintain the fiber, the interconnects, and the service in general. They also must repair any problems and chances are they lose money when it's down. So, you get a lot for the service fee.

Who benefits from the fiber growth?

For one, it's the fiber manufacturers like Corning. They will be on the edge of providing so much fiber to the suppliers that need to lay it out. Not only for the long runs outside but the shorter runs in building and at tower sites. The remote sites are already relying more and more on fiber and less and less on copper. They still need the CAT 5 and CAT 6 jumpers to build the sites, but you see more and more fiber jumpers.

There are vendors that supply the jumpers like RFS, Amphenol, and CommScope that intend to catch the wave. They will supply the material and build jumpers that match what the carriers need. This is a niche that needs to be filled. There will also be the vendors that supply the outside runs of fiber.

The installers, all the construction teams that run fiber and install it. Not only outside, but the companies that install in cities will make a mint, they will be doing very well. There is an ongoing demand for fiber everywhere. While I talk a lot about small cell and macro sites, CRAN will be a key driver as well. All the Wi-Fi providers and if the FCC ever gets off its ass to release the CBRS, then that will be a driver as well. They all need fiber.

What about indoors? The indoor installers will be in demand as well. They will be called on to increase fiber runs in building. They will be harder to track because these are generally electrical workers that added data cabling to their portfolio.

The backhaul providers are in a key position to make money. I am talking about Crown Castle and Extenet. Companies like that are on the cutting edge of growth. That's why Verizon and AT&T have been buying fiber providers; they want to be in the driver's seat. It is to their advantage to use these groups for their own rollouts and if possible, make money off their competitors.

Wireless overview:

Wireless seems like such a good alternative, doesn't it? The problem has been that the growth of the broadband has outpaced the spectrum and equipment. There is a limitation in what the equipment can squeeze through the existing spectrum.

Spectrum:

It may help for you to understand how spectrum affects the wireless systems. I am concentrating on the USA for the purposes of this article.

- ISM band – This is spectrum in the 900MHz, 2.4GHz 5.25GHz, 5.8GHz and 60GHz bands. You already know that Wi-Fi is dominantly in the 2.4GHz and 5.8GHz ranges. It works very well and it completely unlicensed. This has also been used as unlicensed backhaul. In the beginning, it was great because there were few users. However, it didn't take long for many to jump on the bandwagon and cause interference everywhere. There is nothing they can do because it is license free, meaning anyone go jump on anywhere.
- ISM 60GHz band – I am breaking this out separately because there is a lot of

spectrum here and we may be able to take advantage of the 57GHz to 64GHz spectrum for fixed wireless. The advantage is that it's easy to deploy and quick to turn up and unless someone is right next to you there will be little interference. The downside is that it is very limited in distance. Very limited, so there are no long hops in this, just a few hundred feet if that. But the throughput is amazing; some say 4.6Gbps. WOW! However, very short range. The good news is that it's point to multipoint, PMP, meaning it can serve several users from one hotspot. Remember, this is unlicensed, but interference isn't as much of an issue.

- E band – 71 to 76GHz and 81 to 86GHz bands are used here. It is a great way to connect these links, getting over 1Gbps per link, maybe more. These links are great to get a short connection up quickly and easily. It almost is like wireless fiber. The units are generally small and easy to put in. They run well, and they are reliable. The downside is that they can fade with rain.
- 2.5GHz – currently Sprint has over 100MHz of this spectrum. There is more

available that the federal government is trying to pull back to use. This is a great spectrum for 5G mobility and fixed wireless. This spectrum is TDD, Time Division Duplex, which basically means that all 100MHz of spectrum is in one big batch. This means that you can control the amount of bandwidth for uplink and downlink. It is a cleaner system for data communications than FDD, Frequency Division Duplex, which has dedicated uplink and downlink spectrum. If Sprint can utilize this spectrum in 5G and massive MIMO, this will be an excellent alternative to running a cable to any building. The has been technology in the past. Now it's catching up and this spectrum will be the prime spectrum going forward for mobility but also for FWA applications.

- 24.25GHz to 27.50GHz – this is a licensed 5G band that many of the carriers have been buying up. It was reserved for something else, not it's going to be the 5G spectrum of choice for FWA. There is a lot of spectrum and it's great for rolling out the new technology of 5G. It is limited in distance and it's not the greatest for

mobility, but we're looking at FWA, so that doesn't matter right now. Verizon, AT&T, and T-Mobile are all looking to use this to compete with fixed broadband models. It will be interesting to see how they all roll it out.
- CBRS – 3.5GHz band is a newer lightly licensed band. It was used for Wi-Max, which didn't take off, so the FCC pulled it back to be used for LTE. The problem is that they are having a hard time determining how the spectrum will be sued. Specifically, how much will be lightly licensed for private systems and how much for the national carriers. The FCC could make a lot of money if they give more to the carriers, which it looks like that's what they're doing. The problem is the smaller ecosystem of private lightly licensed LTE systems will suffer. The great news is that there is something called LAA, Licensed Assist Access, that will aggregate the licensed carrier with the lightly licensed carrier to make the pipe look huge. This means more broadband no matter who owns the spectrum if they can work together.
- 37GHz to 40GHz – this is a licensed band that is not being utilized, yet.

Spectrum matters for several reasons. The amount of spectrum will determine how much throughput you can get. The way the spectrum is allocated by the FCC helps determine how it is used and if it is PTP or PMP. The format used on it, LTE or 5G also determines the throughput. If someone can license it for a region or for a small area also influences the use. All of this is determined by what spectrum it's in and how it is allocated to the end user.

There were limitations on the equipment in the higher bands, but the need has become so great that OEMs are scrambling to create new radios for those bands. That is if it's a format approved by the FCC.

Spectrum for 5G Fixed Wireless

Let's talk about 5G and fixed wireless. Fixed Wireless is where we put a radio up and shoot it to a building or a kiosk or a small cell or even a macro cell. If you're in the wireless deployment business, then you would call this point to point, PTP, or Point to Multi-Point, PTMP. We used to call these microwave hops, but in this case, it is broadband to a specific facility. While this has been done for a while, not it's going to be a viable competitor to ISPs and Cable companies. Wireless is taking over, and we have a shot to change the world here. Not just the carriers, but the small businesses who want to become ISPs really have a shot to provide real bandwidth to business and home customers. WOW! Can you feel it, a new era is rising in wireless broadband!

First, let's look at the 5G spectrum. I'm not sure if any of you saw it, but the 5G Americas group put together a great sheet on the 5G spectrum. I have the link so go ahead and download it.

http://www.5gamericas.org/files/9114/9324/1786/5GA_5G_Spectrum_Recommendations_2017_FINAL.pdf

Look at the new bands that the FCC is proposing to use:

- 24 GHz bands: 24.25-24.45 GHz and 25.05-25.25 GHz
- Local Multipoint Distribution Service (LMDS) band: 27.5-28.35 GHz, 29.1-29.25 GHz, and 31-31.3 GHz
- 39 GHz band: 38.6-40 GHz • 37/42 GHz bands: 37.0-38.6 GHz and 42.0-42.5 GHz
- 60 GHz bands: 57-64 GHz and 64-71 GHz (extension)
- 70/80 GHz bands: 71-76 GHz, 81-86 GHz, 92-95 GHz

If you think this is just ridiculous because you remember that this is merely a point to point short haul solution, at least if you're in the business and been around, you might be surprised that both Verizon and AT&T are bid on Straight Path for their 38GHz licenses, http://www.phonearena.com/news/AT-T-outbid-for-Straight-Path-by-mystery-firm-rumored-to-be-Verizon_id93451 to gain that particular spectrum. Verizon won!

Don't forget about the Verizon XO deal, http://www.telecompetitor.com/in-pursuit-of-5g-spectrum-verizon-xo-purchase-closes/ where Verizon wants to lease the 28 to 31 GHz and 39 GHz spectrum.

Even T-Mobile is getting in on the act, over a year ago they tested 5G on 38GHz, http://www.fiercewireless.com/tech/t-mobile-files-to-conduct-5g-tests-at-28-38-ghz, to see how viable it is.

I don't see this spectrum as a mobile solution, but more of a fixed solution. That could change. This is going to be a thorn in the side of the cable and ISP business model. Why? It's a new competitor that will have the reach. They have a large customer base. They know how to steal those customers. Remember, fewer millennials are watching traditional TV, they watch on demand as most of you do. Don't deny it, do you really sit down and watch a show at the designated time or do you watch it on Amazon or Netflix or Zulu or with your DVR?

I watch it on my DVR, but also on Amazon. I don't watch much at the time the show airs except live sports. There is the demand for live sports, but I can watch sports on a device as well.

So, the push for massive broadband is coming. Those crappy TV packages that the cable companies are pushing will slowly fade away. Seriously, why can't we get the channels we want and why do they always change the lineup which cancels the one channel we watch? Don't you hate paying a lot of money for something that you don't really want? The cable companies are not changing fast enough in my opinion. However, they do have great internet speeds to the home. I won't deny that. Some markets are better than others and reliability varies.

The wireless carriers are going to have to make it cost effective, so all they need to do is come in lower, about 25%, on price to compete. They can't come in at the same price because cable is very reliable. Wireless will need time to get there. They need to work out the bugs. Then they will do what they always do, slowly raise the price until you leave.

Of course, CBRS will also be a game changer for those businesses that don't need 100Mbps. It will allow us to do more in the rural areas. I love this spectrum because it will be open to more than just the big bad carriers who rule the spectrum. Disruptors have a chance to create something great. This may be the most valuable of all because it may not require LOS, line of sight, as shown here, http://www.telecompetitor.com/fixed-lte-in-cbrs-band-not-expected-to-require-line-of-sight-for-fixed-wireless/ for the connections to be made. This opens new doors for connectivity. It's real and exciting! Hey don't take my word for it, ask Google, http://www.rcrwireless.com/20161117/carriers/google-sees-cbrs-spectrum-band-key-5g-new-model-industry-tag2, and they will vouch for this.

To be fair, the 5G Americas Spectrum document that I referenced above also has a quick blurb in it about CBRS, and I quote "Other bands of interest, From the point of view of global harmonization in the 3 to 5 GHz range as the main mid-range spectrum target for 5G, interests have been expressed in use of this range for 5G in the United States. This could potentially include current CBRS band (3.55-3.7 GHz) and beyond (e.g., up to 4.2 GHz)." The CBRS will play a large part because the carrier doesn't want to deploy small cells everywhere, in fact, they are going to let that up to the business owners and landlords to do. They won't admit this, but I think they are looking for a neutral host solution and CBRS is a great solution! Licensed and protected and it could potentially have multiple carriers on one small cell. A multi-carrier small cell solution. If you think this is crazy, have you ever heard of Wi-Fi? Does it discriminate based on a carrier in your home? NOPE! It just connects, so this will be a stepped-up version of that where it will connect, but it may discriminate based on your carrier. Just apply the proper ID, or ESSID to connect. Amazing!

System distinctions:

There can be several types of links. There used to be point to point, PTP, links that only offered a direct pipe, like fiber, from one spot to another. This was in use for decades for carriers and public safety. It served everyone well until the broadband explosion in 2010. Then those systems slowly became outdated.

Then came the point to multipoint, PMP, systems that allowed one hotspot to reach several remote nodes. This is exactly how your Wi-Fi hotspot works. Same as the carrier's mobile systems. They talk to many from one access point.

The big difference between the two, other than one talks to one versus one talks to many, comes down to antenna properties. The PTP system can concentrate all the RF into one beam, giving the RF the strength to go farther and penetrate more. Whilst the PMP loses power by widening the beam width to allow a broader number of users to talk to it at the same time.

Also, the PMP system has a longer delay due to processing multiple users versus just one.

Costs in wireless:
The costs are still here; it is not free. While the equipment may be expensive to put in, it's not the most expensive part. It's an upfront cost, and then it's in.

While there is usually not a monthly fee, there could be maintenance and site rental. Monthly rent could be higher if the equipment is large.

However, Sprint has done a great job using UE backhaul, it is a backhaul over their 2.5GHz TDD system that is already coming from a macro site, and the end unit at the site is small. Not much space on the asset and it has enough throughput for a small cell. While it's not CRAN fronthaul solution, it does quite well for a small cell.

Resources:
- https://wade4wireless.com/2018/02/26/how-do-we-get-more-backhaul-bandwidth/
- https://wade4wireless.com/2015/10/19/what-is-lte-ue-backhaul/
- https://wade4wireless.com/2015/03/13/smart-small-cell-deployment-backhaul-efficiencies/
- https://wade4wireless.com/2018/02/05/laa-cbrs-lte-u-are-5g-building-blocks/
- https://wade4wireless.com/2018/01/29/about-massive-mimo-beamforming/
- https://wade4wireless.com/2018/01/01/the-art-of-solutioning-estimating-for-profit/
- https://www.corning.com/media/worldwide/coc/documents/Fiber/FSwin15pp3334.pdf

- https://en.wikipedia.org/wiki/Backhaul_(telecommunications)
- https://www.slideshare.net/3G4GLtd/different-types-of-backhaul
- How Comcast and AT&T stopped Google Fiber, https://arstechnica.com/tech-policy/2018/01/att-and-comcast-finalize-court-victory-over-nashville-and-google-fiber/
- https://www.pcmag.com/encyclopedia/term/45467/ism-band
- https://arstechnica.com/gadgets/2016/12/802-11ad-wifi-guide-review/
- https://www.e-band.com/70-80-GHz-Overview
- http://www.e-band.com/get.php?f.850
- http://www.electronicdesign.com/communications/what-s-difference-between-fdd-and-tdd
- https://www.cept.org/ecc/topics/spectrum-for-wireless-broadband-5g
- https://www.qualcomm.com/media/documents/files/spectrum-for-4g-and-5g.pdf
- https://www.wirelessweek.com/news/2016/07/fcc-unanimously-opens-nearly-11-ghz-spectrum-5g

- What is 5G?
  http://spectrum.ieee.org/video/telecom/wireless/everything-you-need-to-know-about-5g

- http://www.5gamericas.org/files/9114/9324/1786/5GA_5G_Spectrum_Recommendations_2017_FINAL.pdf
- http://www.5gamericas.org/files/2414/1323/5229/4G_Americas_Spectrum_Sharing_-_FINAL_Oct_2014.pdf

- What is 5G?
  http://spectrum.ieee.org/video/telecom/wireless/everything-you-need-to-know-about-5g

- http://www.5gamericas.org/files/9114/9324/1786/5GA_5G_Spectrum_Recommendations_2017_FINAL.pdf
- http://www.5gamericas.org/files/2414/1323/5229/4G_Americas_Spectrum_Sharing_-_FINAL_Oct_2014.pdf

What is Fixed Wireless?

Think of your internet access at your home. Many of you have cable modems or fiber or DSL or satellite. Fiber would be FTTH, which you may call Verizon FIOS or AT&T U-Verse. The wired solutions are expensive for the larger companies to deploy, just ask Google who thought they could do it for less money but learned the hard way that physical attachment to poles takes more than just goodwill to the city. I talked to my friend in Nashville where the poles had rights of refusal by AT&T and the local cable companies that did NOT want Google to play in their neighborhoods. It did not matter what the city said; whoever had rights to the poles had the final say!

That is where the wireless option looks attractive for many reasons. 5G technology, like cmwave, mmwave, and CBRS can help make this happen. We still need fiber, that part is crucial, but we do not need to run it to every home. There is an opportunity to build out FWA to the home using 3.5Ghz or 28GHz, all depending on the location and distance to the BTS.

By the way, this has been done before with microwave connecting buildings for telecom services. It's not new. It is just cheaper and faster and better. We are an all-IP network now which makes the transport invisible to the network. Now we have a spectrum that we can use with better technology. We can shape the broadband rollout to improve the broadband infrastructure in a profound way. The technology has arrived.

Can we get more spectrum? It looks like the US FCC took the first steps, they have opened 28GHz (27.5–28.35GHz), 37GHz (37–38.6GHz), and 39GHz (38.6–40GHz) for this purpose. Also, 7GHz of the unlicensed spectrum from 64–71GHz. If you remember, some of these bands were used in the past to deliver point to point, PTP, microwave for building access. Now that the equipment is changing and becoming more cost-effective, it can be used in new applications. MIMO antennas and systems are also helping the cause. Multipoint radios are becoming more and more available. Technology has come a long way!

I am looking forward to seeing the first mass fixed wireless rollout. If we can get broadband to the homes without cables or fiber running through the house, how cool would that be? If small businesses could have broadband in their stores and homes without waiting for fiber to be deployed, life would improve for everyone.

Broadband is the new utility! If we could have a unit that we could put on a window facing one antenna outside and have the Wi-Fi or LTE-U or CBRS inside, life would be grand! I see this is coming sooner than later.

The carriers are pushing to get fixed wireless out to the public. They have been trying to work with several technologies. It looks like LTE will be the foundation of the format. It could be mmwave or spectrum they have for LTE today. The carriers will tell you that this is 5G, but it has more to do with advanced LTE being able to push the limits using carrier aggregation in the current spectrum and making new spectrum multipoint. Carrier aggregation and MIMO makes larger wireless broadband realistic. Sprint is in a great position with all the 2.5GHz spectrum they should pull this off quickly. If only they would spend the money to do it.

With mmwave, we have very large bands. The great news is that it could be deployed quickly. I think it will be lightly licensed because the coverage area is so small. I also think it could be the solution to getting massive amounts of spectrum to building in a short time. The current systems are point to point, but they are rolling out multipoint systems. There is an [article in Gigabit Wireless that helps to explain](#)

more about mmwave and the multipoint technology in that band.

A 1Gbps links will make it possible to run 100Mbps to multiple homes from one cell, be it a small cell or a Macro. Macros and Wi-Fi can do that now. All the carriers are promising this.

- http://bgr.com/2017/01/04/is-5g-faster-t-mobile-att-sprint-verizon/ Where he says AT&T, T-Mobile, and Sprint all promised 1Gbps in 2017.
- http://www.rcrwireless.com/20160726/carriers/verizon-5g-trials-seeing-1-gbps-speed-expected-lower-costs-tag2 where they explain that Verizon already sees 1Gbps.
- http://www.fiercewireless.com/wireless/sprint-cto-hints-at-1-gbps-class-speed-2017-challenge-to-t-mobile where Sprint promises 1Gbps.
- T-Mobile hits nearly 1Gbps in the lab, https://www.wirelessweek.com/blog/2016/12/watch-t-mobile-hits-nearly-1-gbps-lte-lab-test

I believe that we will see a fixed wireless solution very soon. I believe that 100Mbps to the house via a wireless link is very realistic. This will be a game changer that will have a dramatic effect on our daily lives.

It has to compete with cable. With my cable modem, I feel I get good speed. I tested it and got 67.3Mbps down and 11.9Mbps up. I am happy with this at home, today, which I show from Google's internet speed test on Comcast, shown below. Way to go Comcast! Can Verizon, T-Mobile, and AT&T compete?

If I get fixed wireless, will it work this well? I hope so. As you can see, it must give about 100Mbps to each home. The cable company can do this today. Verizon and AT&T both offer this over fiber. If they do it with LTE, then they can compete.

The question now is, is it cost-effective to use FWA over other technologies? The installation and setup will determine that. That is why I say KISS, Keep It Simple Stupid! The key is to make the installation so easy that anyone can do it. When I say anyone, I mean the end user. No one wants to pay $1,000 for installation. Most companies may do this for 100Mbps of service, but at home. Most people bitch about overpaying because they don't get it. The cable company to come out and install their equipment for a few hundred dollars. I know I do. Especially when you are paying over a hundred a month for service. Home users want value at an affordable price.

Home businesses need reliability, so make it very reliable. They will pay more for reliability. Price and reliability which will be determined by competition, which is one thing that most cable companies do not have right now. Seriously, whoever has a connection to your home is the winner, and cable modems are way faster than DSL. Will that change with FWA? Will cable be competing with the wireless carriers? Of course.

If they can make the installation simple, I would be happy. Would you?

Have the outdoor wireless connect to something simple and effective indoors. Let the people see the signal level for the outdoor connection, like DISH used to do, and make it broadcast Wi-Fi inside and offer wired Ethernet. If we can get this, life is great. We can connect our own router or use what they give us.

Fixed Wireless Overview

What is the future of fixed wireless? Chances are it will take off, in fact, the major carriers are counting on it!

Is 5G mostly fixed wireless? Some of it will be, along with IOT, massive broadband, augmented reality, and surprising mobility.

Will fixed wireless replace fiber to the home? The carriers are hoping it does because of the cost-effectiveness and the ease of installation.

Will fixed wireless replace cable modems? Again, the carriers are betting on this, and the cable companies know this, they know that can do something and finally enter the wireless arena, for real this time.

Fixed wireless access, FWA, is going to be a game-changer in so many ways. It is going to be part of the 5G network slicing that we have all heard about. There is a spectrum, like CBRS, mmwave, and CMwave that will make it or break it. Verizon, AT&T, T-Mobile, and Sprint already are testing this on MIMO antennas. They are counting on a new income stream. The question will be, what is the ease of install to the end user, the consumer, you and me? Do we still need someone to come out and wire up the house? Do we need someone like the DISH network guys to put an antenna on the roof? Alternatively, maybe, can we just put a unit in the window that could receive the licensed or lightly licensed signal then transmit Wi-Fi in the home? Wouldn't that be cool? Just like the wireless modems we used to know only on steroids giving us speeds of over 50Mbps and up. That is the dream, right? Hopefully, it will work regardless of bad weather or good, power or no power (UPS backup) that businesses and homes have massive broadband with under an hour of setup.

What is fixed wireless? Think of your internet access at your home. Many of you have cable modems; some have fiber to the home, FTTH, which you may call Verizon FIOS or AT&T U-Verse. The wired solutions are expensive for the larger companies to deploy, just ask Google who thought they could do it for less money but learned the hard way that physical attachment to poles takes more than just goodwill to the city. I talked to my friend in Nashville where the poles had rights of refusal by AT&T and the local cable companies that did NOT want Google to play in their neighborhoods. It did not matter what the city said; whoever had rights to the poles had the final say!

That is where the wireless option looks so attractive for so many reasons and 5G technology, like cmwave, mmwave, and CBRS can help make this happen. We still need fiber, that part is crucial, but we do not need to run it to every home. There is an opportunity to build out FWA to the home using 3.5Ghz or 28GHz, all depending on the location and distance to the BTS.

By the way, this has been done before with microwave connecting buildings for telecom services; this is not new. It is just that now we have a way to get it to each business and we are an all-IP network now. This technology is available today and being done by point to point microwave as well as multipoint systems. It is just now we have a spectrum that we can use with newer and better technology. We can shape the broadband rollout to improve the broadband infrastructure in a profound way. The technology has arrived.

Can we get more spectrum? It looks like the US FCC took the first steps, they have opened 28GHz (27.5–28.35GHz), 37GHz (37–38.6GHz), and 39GHz (38.6–40GHz). It is something that we can use, I hope. Also, 7GHz of the unlicensed spectrum from 64–71GHz. If you remember, some of these bands were used in the past to deliver point to point, PTP, microwave for building access. Now that the equipment is changing and becoming more cost-effective, it can be used in new applications. MIMO antennas and systems are also helping the cause. Technology has come a long way!

I am looking forward to having fixed wireless rollout. If we can get broadband to the homes without cables or fiber running through the house, how cool would that be? If small businesses could have broadband in their stores and homes without waiting for fiber to be deployed, how great would that be? If we could only have a unit that we could put on a window facing one antenna outside and have the Wi-Fi inside, life would be grand! I think this is coming.

The carriers are pushing to get fixed wireless out to the public. They have been trying to work with several technologies. Whatever they work with it looks like LTE will be the foundation of the format. It could be mmwave or spectrum they have for LTE today. The carriers will tell you that this is 5G, but it has more to do with LTE being able to push the limits using carrier aggregation in the current spectrum and making new spectrum multipoint. Carrier aggregation makes that look realistic. I think Sprint is in a great position with all the 2.5GHz spectrum they have to pull this off quickly. If only they would spend the money to do it.

With mmwave, we have very large bands. The great news is that it could be deployed quickly. I think it will be lightly licensed because the coverage area is so small. I also think it could be the solution to getting large amounts of spectrum to building in a short time. The current systems are a point to point, but they are rolling out multipoint systems. There is an article in Gigabit Wireless that helps to explain more about mmwave and the multipoint technology in that band.

I believe that 1Gbps links will make it possible to run 100Mbps to multiple homes from one cell, be it a small cell or a Macro. Although only a Macro can do that now, it must improve. All the carriers are promising this.

- http://bgr.com/2017/01/04/is-5g-faster-t-mobile-att-sprint-verizon/
  Where he says AT&T, T-Mobile, and Sprint all promised 1Gbps in 2017.
- http://www.rcrwireless.com/20160726/carriers/verizon-5g-trials-seeing-1-gbps-speed-expected-lower-costs-tag2
  where they explain that Verizon already sees 1Gbps.
- http://www.fiercewireless.com/wireless/sprint-cto-hints-at-1-gbps-class-speed-2017-challenge-to-t-mobile
  where Sprint promises 1Gbps.
- T-Mobile hits nearly 1Gbpa in the lab, https://www.wirelessweek.com/blog/2

016/12/watch-t-mobile-hits-nearly-1-gbps-lte-lab-test

I believe that we will see a fixed wireless solution very soon. I believe that 100Mbps to the house via a wireless link is very realistic. This will be a game changer that will have a dramatic effect on our daily lives. With my cable modem, I feel I get pretty good speed, today I tested it and got 67.3Mbps down and 11.9Mbps up. I am happy with this at home, today, which I show from Google's internet speed test on

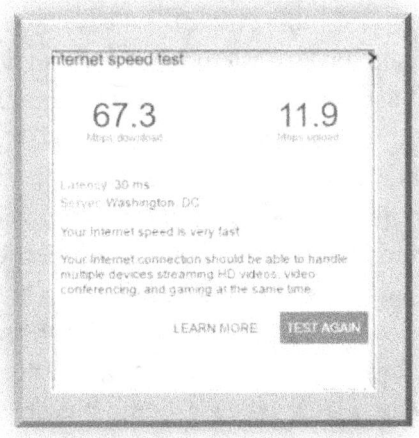

Comcast, shown below. Way to go Comcast!

If I get fixed wireless, will it work this well? I hope so. As you can see, it must give about

100Mbps to each home. The cable company can do this today and more. Verizon and AT&T both offer this over fiber and more. If they do it with LTE, I see TDD working better the FDD so that they can proactively balance the upload and download speeds. That is why Sprint has a prime spectrum with their 2.5GHz band. This band travels well and would work great as a fixed wireless platform.

The question now is, is it cost-effective to use FWA over other technologies? The installation and setup will determine that. That is why I say KISS, Keep It Simple Stupid! The key is to make the installation so easy that anyone can do it, as the end user. If someone must pay $1,000 for installation, it may not be cost-effective. Most companies may do this for 100Mbps of service, but at home, people bitch about overpaying the cable company to come out and install their equipment for a few hundred dollars. I know I do. Especially when you are paying hundreds a month for service. Home users want value at an affordable price. However, home businesses rely on reliability, so make it very reliable. Price and reliability which will be determined by competition, which is one thing that the cable companies do not have right now. Seriously, whoever has a connection to your home is the winner, and cable modems are way faster than

DSL. Will that change with FWA? Will cable be competing with the wireless carriers? Of course.

If they can make the installation simple, easier than hooking up my phone, I would be happy. I do not see why not. Have the outdoor wireless connect to something simple and effective indoors. Let the people see the signal level for the outdoor connection, like DISH used to do, and make it broadcast Wi-Fi inside. Preferable Wave 2 with the ability to connect an indoor router via wired Ethernet. Then life is great!

Will this be easy for the carrier or service provider to do, not really? However, would it be easy for the cable companies to roll out, yes? They have the infrastructure to make this happen. They could deploy the radios efficiently and quickly. They have the workforce and the structure to handle business and residential. If only they had the spectrum. If only the cable companies would move into the wireless realm. They would be a force to be reckoned with. They already have a huge customer base, and they have the core and the support centers. I think that cable companies are positioned well. Will they roll something out? If they can get in on the CBRS or the mmwave or the

I just heard an interview with John Legere where he explains, (I am paraphrasing) how companies are identified by their infrastructure, wireless or cable, and the end user could care less. I agree with this. I think that people just want broadband when they need it, whether it is home or on their device or in a coffee shop. I agree with Legere when he explains how mobile is taking over and that people just want to have a great connection. He has been on fire lately because T-Mobile has had a kick-ass year and he will not stop. He turned T-Mobile into a player, putting Sprint behind him and making AT&T sweat.

I want to congratulate John Legere and T-Mobile for winning a ton of 600MHz spectrum in the recent auction, great job T-Mobile for getting national coverage after all this time. He says, "Little Ole T-Mobile," but they are not little anymore, in any way.

For more look at all the John Leger interviews listed below.

- https://youtu.be/LVQuD_fTpcE
- https://youtu.be/LgnMKD0xmd8
- https://youtu.be/T4MK_t1z-gw

I think that he makes a great point. I think that the internet providers will be listed as providers and as companies like Google will be media providers. I think that AT&T is trying to play both sides. There are going to be providers of the service and providers of the content. Who is going to win in the upcoming battle? I am not sure, but we need to stop looking at cable companies and carriers for service their specific audiences, and they will start service everyone. Barriers are coming down, and the gloves will come off.

Will cable companies merge with carriers to remain competitive? Probably, look at Comcast working with Verizon and AT&T taking over DISH. Competition is rising. Comcast has the money to start their wireless system or take over a player like T-Mobile, but will they spend the money? They have not so far, but the playing field is changing, and Comcast sees the writing on the wall. It is time to make something great happen!

As a final note, and a way for me to bring smart cities into this. I believe that all smart cities want competition in broadband, they want the service everywhere in their cities, so the FWA will make that option a reality of the carriers build the entire city. All areas of the cities need to be served, not just the business districts or the upscale neighborhoods. I get that the carrier wants payback, but we need to blanket cities to give everyone an equal opportunity!

This is making broadband the new infrastructure backbone of America and giving us all an opportunity to play. Let's make something great happen!

Who will provide broadband?

Will cable companies merge with carriers to remain competitive? Probably, look at Comcast working with Verizon and AT&T taking over DISH. Comcast and Charter are looking at buying Sprint. Competition is rising. Comcast has the money to start their wireless system or take over a player like T-Mobile, but will they spend the money? They have not so far, but the playing field is changing, and Comcast sees the writing on the wall. Is it time for Comcast to make something great?

When you look at what AT&T and Verizon are trying to do with fixed wireless testing, you see 2 companies that want to replace the FTTH models. They have every intention of using this for consumers to compete directly with the cable companies as well as provide new services to the mobile customers. However, let's take it a step farther. If they can provide their own backhaul as well as businesses an amazing experience, it becomes a win-win. Why not? They could provide the backhaul to small cells and mini macro sites. They could be the fronthaul for the CRAN sites. The system would make the carrier its own customer. The money could circulate inside its own business without the expense of running fiber everywhere! How amazing to have something that could be used for consumers, businesses, and your own business. It all makes sense

While the other carriers in the US, T-Mobile, and Sprint, don't seem to have the commitment to fixed wireless they do intend to get their mobile systems to provide limited broadband to any device. Why shouldn't they all compete for the broadband business?

Why does Fixed Wireless Matter for City Growth?

Why? Because cities are going to want alternatives to running fiber on poles and underground. The dig once the policy is going to be enforced more and more throughout the USA. The poles are going to be a point of contention among competitors. The access rights and permitting battles will heat up until we find a resolution. The FCC is working to streamline small cell deployments, and the cities are realizing that they must lay out the requirements for proper installation. If things go as planned, the fixed wireless base stations should be a lot like small cells with batteries. I believe that power is going to be the issue because if the power goes out people still want Internet access. So, this issue needs to be resolved.

Cities are aware that they need broadband service for small businesses and for kiosks as well as food trucks. Today they rely on smartphones, but the demand will grow. As demand grows we come up with new solutions. If you go to a city, you will see stands that sell newspapers, hot dogs, and T-Shirts almost on every busy street. These will all be connected, some will be automated, and all will be within coverage of a wireless signal, either mobile or fixed. Today, many need landlines and smartphones to do business. Wireless is a game changer. Most people think that broadband only serves online business, but the brick and mortar need broadband as much as anyone. Let's not forget services that have offices, like lawyers, realtors, city workers, lessors, and so on. All in a city with offices that need to be there to do business. As their business grows, revenue grows, the city gains more in tax revenue. It's a win-win all around. That's why quick and economical broadband connection is vital. Let's face it, many businesses use that today, in the form of a smartphone and some type of automated payment systems like Apple Pay or Square.

Fixed wireless is an alternative to mobile and wired that could help businesses grown.

You probably know this, but cities rely on small business growth to maintain city growth. Most small and medium businesses will rely on broadband access to grow. Small and medium business growth will make or break city growth. In some cities, they rely on the citizens to make this happen. Some rely on a big company, like Comcast, to make this happen. I think they would be smart to allow all businesses to grow.

However, it's not always up to them; they must deal with the incumbents taking over and keeping competitors out. I brought up Nashville and their quest to get Google Fiber in, but AT&T and Comcast fought it with everything they had, and they won.

Fixed Wireless could be a key to Growth for Urban and Sub Urban areas.

Fixed wireless will be an alternative to wired internet access that could increase competition overnight. New wireless spectrum and technology is opening the door for more and more broadband, enough that we can see an exponential gain in broadband to the buildings without wires. We all lover fiber, but the cost is prohibitive to most small businesses. Think about getting internet to a home business; they may not get fiber due to the cost. Let's face it, running fiber everywhere is not cost effective. FWA may be an answer to cable modems and DSL. Why do you think so many small businesses rely on their smartphone, they know it's effective, mobile, and works? It's better than paying the cable company even more money. Fixed may allow the customer to connect the last 200 feet from the nearest fiber.

Many people rely on LTE for connections; they understand that their smartphones can do a lot for business. Beyond email, smartphones are used for business payments using credit card interfaces for purchases. Most purchases are tracked via an app or on the web. Mobile devices have improved mobile businesses.

The small office space needs this support. We are almost there with wireless devices working so well in buildings. Cost prevented this before, but now it's becoming cost-effective with new and improved and smaller technology.

Adding more wireless connections will help the city infrastructure. Wireless will reduce the amount of fiber and cables run throughout a city or building. Today, one of the largest expenses in getting broadband is the cost of fiber. So, the fiber can be run to the wireless hub or router so that the 200 feet will be wireless for many small to medium businesses. This will work if the cost models allow it.

Think about the process of getting fiber. The permitting and planning takes time. First, getting the fiber to a location will need to be planned because you need to know where it is coming from, a demarcation, and how to get it there, pole by pole or underground. Then, digging or mounting to poles requires permits from the city or local municipality. Poles require leases, more time and more money. There is more work along with the time it takes to do all of this. If you do a wireless hub, you install the fiber once, install the radio once, done. The customers then will add wireless interfaces or modems where possible in a matter of hours, not months.

Wireless saves so much on the last 200 feet! Cities deal with problems when running fiber everywhere, like traffic problems when streets are closed while they are being dug up to bury the fiber. The streets are never the same, so they may need to be resurfaced. This is very expensive. Why do you think small cells were so delayed in many cities? It was getting the fiber to each location and the aesthetics of the mount, in other words, backhaul. These obstacles are being overcome. If underground, then streets generally get dug up if there are no tunnels or underground conduits to run new fiber in. You could go overhead if poles are available, and that's what many companies do, overhead if another company doesn't have exclusive rights to that pole. This has been an issue; Google fiber had to deal with incumbent almost everywhere they tried to roll out. The incumbents hate competitors. So, they will do anything they can to stop them. AT&T and Comcast and Cox and CenturyLink do what they can to keep anyone else out. Don't get mad at them, that's just good business.

Most cities today have a dig once policy, but what about when someone new comes to town? Then they may want to dig again, or they are forced to use a competitor's fiber. We also should think about how to minimize the digs and to solve the problems of adding fiber to existing poles. Pole leases and permitting can get complicated. So, what do we do? We go wireless, to be more specific, fixed wireless that has a high bandwidth from a nearby spot. This will extend the fiber that's there to new locations. While the 60GHz spectrum would be a good solution, Verizon and AT&T invested heavily in the 28GHz spectrum to improve 5G specifically for fixed wireless. The technology is here and the costs reasonable. Deployment is getting easier and easier.

This will be called 5G technology, but the concept has been around for years with a point to point microwave. There were companies that did this in the early 2000s with success, but the costs were excessive. Fiber costs were dropping and became the connection of choice when it was widely deployed to a larger building in larger cities.

Now we want broadband everywhere in every business in every office, large or small. We need to get creative to deploy broadband everywhere. Every day we want that cost to come down.

Will fixed wireless replace fiber to the home? In the city, it could, cities are dense and could work with this model if the wireless companies can compete with the local cable company. The wireless carriers are hoping it does. They will rely on lower costs and ease of installation and mobility to make that happen. If they can get the 28GHz to work properly here in the USA and get a device that anyone can install, they will put a serious dent in the cable company's business plans. Suddenly, competition will be running rampant, if they can deploy. The business model must make sense.

Will fixed wireless replace cable modems? Again, the carriers are betting on this. I am sure the cable companies know this and intend to get more and more into wireless delivery. In my opinion, the key is easy and painless installation and commissioning. Something that is not so easy for the cable company.

Fixed wireless access, FWA, is going to be a game-changer in so many ways. It is going to be part of the 5G network slicing that we have all heard about. There is a spectrum, like CBRS, mmwave, and cmwave that will make it or break it. Verizon, AT&T, T-Mobile, and Sprint already are testing this on MIMO antennas. They are counting on new income streams. The question will be, what is the ease of install to the end user, the consumer, you and me? Do we still need someone to come out and wire up the house? I hope not! Do we need someone like the DISH TV guys to put an antenna on the roof? Hopefully not in the city. Alternatively, maybe, can we just put a unit in the window that could receive the licensed or lightly licensed signal then transmit Wi-Fi in the home? Wouldn't that be cool? Just like the wireless modems, we used to know; now they are on steroids giving us speeds of 50Mbps and up. That is the dream right, bad weather or good, power or no power (UPS backup) that businesses and homes have massive broadband that we can set up in under an hour and take with us if we move.

Unseen Costs of Deploying Fiber and Wireless

The thing about running any type of equipment is more than the cost of the equipment and installation. It's all the other stuff that most magazines and articles won't cover. It's the delays, permitting, acquisition, and approvals needed to be managed. These are necessary evils for any deployment. Just ask any carrier. Therefore they are adamant about fighting local jurisdictions for the small cell costs of permitting and rent.

This is all before you deploy one thing, pay for one piece of equipment, or even design the backhaul. It needs to be worked through, and chances are it will be around 10% to 25% of the cost. If you're talking small cells, then it could be 50% to 75% of the cost.

The rise of Site Acquisition

Let me give you some history, at least from my experience. When we deployed systems years ago, we had to pull permits for structures we built, like towers, poles, and so on. That made sense and you had to get the permission of all the people around the tower as well as the local jurisdiction, like a tower, township, or a county.

However, you didn't need their permission to put stuff on the tower.

When adding to the tower, the owners generally knew what it could hold and what would overload it, so that was not an issue. Cell systems were very simple, usually, an antenna attached to the tower with coax. No big deal.

Then 2 things happened. Local townships wanted to have a permit for everything that attached to a tower or pole. They wanted permitting dollars for each item on the tower. Even though they knew nothing about it, they wanted to have a say in what goes on someone else's tower. If you ever had to pull a permit for a home addition, it's the same thing. While they say it's to make sure they know what's in their jurisdiction, we all are sure it's because they want to tax you more. It's all about the property tax and what they say it's work. I mean, if I finish my basement, how does that affect them in any way? Only by the tax dollars.

Then the cell companies and backhaul companies started adding larger and larger equipment to the towers. It became overwhelming. Everyone started putting radio heads on the towers. They started adding full platforms instead of simple antennas. They added more and more equipment to the top of the tower. Dishes had radio heads on them, and cell antennas had radio heads stacked up behind them. Large radio heads, coax, and multiple runs into an antenna. Having lots of weight on the tower that needs to be certified safe so the tower will not collapse.

So not you have to deal with pulling permits, zoning, and paying the local jurisdiction for access to the tower. Most times it was local, and you used to have to send someone to the local board meeting to get permission. Can you imagine? Verizon and AT&T had to send someone to meet with the local zoning commission on the first Monday of each month hoping to get approval to mount their equipment on a tower? Now, take that times 1,000 and you have the reality of how much it costs to send people around to get permission. By the way, you were lucky if they would get to it or approve it in one meeting. You may have to go to 2 or 3 meetings to make it happen, so that's 2 or 3 months of delay. It's not that bad anymore, but it was like that for years. All the carriers had to deal with this. Now they just need to pull permits. (This is why the carriers have such an issue with local jurisdictions!)

Then you have all the tower issues, like a structural assessment and mount assessments to see if you can make changes to what is on the tower. Any changes need a full structural done. If you plan you can do the assessment with the proposed future equipment.

For the tower side, this added a ton of cost, and it sparked a new industry, the site acquisition industry. They handle all of that, and they earn every penny for all the work and bull shit that they have to put up with. Many jurisdictions see wireless as the enemy, that is until they use their smartphone to call all their friends to complain. They just don't get it.

Wireless Costs

I gave you an example of how the costs of doing tower work is not going away. The site acquisition portion has grown while the carriers did a good job of reducing costs in other areas.

One area was tower climbing. The carriers have reduced that by over 50%. However, did they take too much out? Did they drive it down too far? The carriers have driven down tower services to the point where margins are crap, and they didn't care because until recently they never had to pay anything when a climber falls and gets injured or dies. That all changed in May of 2018 when a jury told AT&T Mobility to pay the family of a fallen climber $30 million. A landmark case and one that will make the carriers think about driving cost out of tower work. Lower cost generally means lower skilled climbers, less training, poor safety gear, and so on. More can be found at http://wirelessestimator.com/articles/2018/att-settles-for-a-record-30-million-to-injured-tower-technicians-family/ about the case.

The other thing that has helped the carriers is automation. While they still rely on drive teams for testing, for the most part, they have automated RF engineering and optimization. This has dramatically reduced the cost and effort put forth.

The labor of any deployment adds up. However, as the work becomes a commodity, the costs go down. So far, the carriers can't replace the tower climber, but they have been offshoring and automating other services to save costs.

Utility power is another cost. If you need power at the site, which you do, running utility to a site costs money. It could also add delays. You need a licensed electrician to work with the power company to make sure you have the power in properly. Then the permitting and inspections that go along with it. It adds delay and cost to any project. You will need power, whether at a pole or in a building or at a tower site. It all adds up.

Before you mount anything:

Before you mount anything, this is generally a 3-month wait. It's similar to fiber. Fiber deployments have all the costs of permitting, permissions, and more. They often run into a dig once policy with cities. This is good for the city because they don't want the streets dug up again and again, but the company running the fiber may have to wait until one or more companies are also running fiber. This adds delays, costs, and competition. If you're deploying first, you have a chance to lock up accounts, but if you must wait for the competition to go alongside you, I am pretty sure they are going to go after the same customers you thought you had locked up. It is a dog eat dog world out there.

Drawbacks of Running Fiber

The thing about running fiber everywhere is not the fiber itself. It is the costs associated with running the fiber. In fact, anything you deploy has high costs. Building towers and adding a radio to a pole has a high cost. While you may think it's the installation and design, that is a small portion. It's all the civil work and the acquisition paperwork that needs to be done to make it happen.

I read a great blog by Michael Dargue at https://blog.cartesian.com/why-is-the-cost-of-ftth-not-falling-faster-five-things-that-dont-follow-moores-law about the costs of running fiber and why it's not getting any cheaper. He compares it to Moore's Law, but in my experience, it's hard to compare services to components. Services don't always get real cheap unless you can do it overseas.

All the things that need to be done like the civil engineering if you go underground and any structural engineering you may need to do on a pole, that needs to be done before anything gets ordered, much less deployed. These costs have yet to be reduced too far. The companies have done a good job with the bidding and reverse auctions, but it still costs money.

The other thing, like tower work, labor. Again, bidding and reverse auctions have driven the costs down, but you still must pay qualified and licensed workers to do the work. In some cases, you have prevailing wage and union fees that must be accounted for. This is not cheap, and chances are they add to the costs. In some cities, you may have to choose from 2 or 3 companies only because of union regulations. While there is competition, you can't bring someone cheaper in from out of town. The costs remain the same regardless.

Permitting is still an issue, not just for the costs but for the delays. In some cases, I have heard of cities letting a company deploying fiber get a blanket permit for an area. This saves time and costs. It is a great idea. All the company needs to do it log where they attach and send in the qualifying documents. This is a great idea for a company deploying fiber across a city. It's an excellent way for the city to get the permit fees but cut down on the labor to process all the paperwork. It seems like a win-win.

If you have a router at a pole, that costs money. Running utility power to a pole is expensive. Even if you need to run your own service in a building, this adds cost. It could add delays. Think about all the permitting and inspections. If you're in a building, then you need to modify the lease to account for power, and you need the permission of the building owner. More delays and costs.

No matter what you deploy:

Costs for utility, permitting, site acquisition, and more all add up. You will need a team of project managers and engineers to manage this. You will need a construction manager to go to the site to verify things are happening per the plan, paperwork, and schedule.

Before companies spend money on any deployment, they have to overcome the hurdles of planning. They need to worry about what and how they deploy, what they will attach to or where, they will bury the fiber. Chances are the end customer will put things out to bid or reverse auction to drive costs down. One thing you learn is that loyalty is disappearing in this new business arena. It's a fight to get the business then a struggle to maintain margin. All the liability is put on the front line, the installers.

They also need to make sure they have all the hardware necessary up front. There are always surprises that jump up and say, "GOTCHA!" when doing any deployment. On a tower it's the mounting hardware and clamps, on a pole attachment is mounting hardware, in the ground it could be going under driveways or a street or an electrical line up ahead. These add delays to the deployment. By the way, don't' forget weather! For all installations, weather can delay everything. Often overlooked and there is nothing you can do. Weather is a risk you must accept. If you're just thinking rain, guess again. It could be a snow storm, extreme cold or heat, hurricanes, or even tornado warnings. They all make the deployment stop. They all add delays, and they are all safety issues.

A good construction manager can make all the difference in the world. They can deal with costs, put in an effective system for change orders and be responsive to problems. They can also plan properly accounting for potential problems. It's all about planning and reacting. They should be able to deal with most problems and have a budget to do so.

Rent Applies to Everything

If you follow me at all, you know that the carriers need to pay the tower companies a lot of money to be on their tower. This is an OpEx cost that they need to deal with. So far, there is no alternative. OK, you said small cells and poles, right? Maybe building tops? Listen, rooftops and building usually have stricter leases and cost more to go into because it is prime real estate. Poles for small cells are an extension of the macro site. Get it, not a cost savings, only additional costs.

However, what about fiber? I mean you bury, it's free right? The rules for right of way allow it to lay there for free? Well, if they attach to any pole or go through someone else's right of way, there is a fee. To attach to a pole, you need to pay something to the pole owner. Albeit, it might be three to ten dollars a pole. It doesn't sound like much, right? The thing is, fiber has to attach to more than one pole, probably hundreds if not thousands to get where it is going. Now we're talking that small amount times a thousand, every month. Get real; there's a cost to most everything. Why do you think to be first matters? You get in and get the business without competition. The cable companies knew how to lock up that business before anyone else could compete. They made deals with the local municipalities and pole owners that would protect them against the competition. They're not stupid. It's business.

When thinking of fiber, there could be monthly costs for access to the internet in specific areas, or maybe building access costs money. Maybe they have routers that need power, another monthly cost. It all adds up.

Summary of costs:
I think to go into a deployment we all look at the hardware and the installation costs. This is easy, hey, five grand for a tower crew and the hardware is about 100 grand for the equipment then maybe another five grand for the groundwork. This is a dream world friend. The

point of this article is to pull you into the real world.

No matter what you deploy, there will be costs that you didn't think of. That's normal. How you plan and react will make the difference.

Broadband Initiatives for Cities

Cities want competition in broadband; they want the service everywhere in their cities, so FWA will make that option a reality across an entire city. To serve most of the city quickly and reliable will be with FWA. This way they don't need to worry about pole rights everywhere. The broadband backbone is key for a city's growth. However, city governments want all the city served, not just the business districts. They can't expect the underserved area to get more business unless they have access to broadband.

Are you curious about some broadband initiatives that are out there? Some broadband case studies that are have rolled out? What broadband initiatives have been successes and failures? What works better, the city owning it or a public-private partnership or private only?

Broadband is the foundation of any smart city. Someone shared the Next Generation Network Connectivity Handbook. It's a publication by Gig.U, and it can be downloaded from http://www.gig-u.org/. Gig.U is a group that encourages the partnerships between cities and universities. They did a great job of putting together this document, published in December of 2016, showing past case studies of gigabit deployments in both wireless and wired. They cover success and failures.

Here is an outline of some of what is in this document, I highly recommend downloading it, after all, it is free!

First off, the make the point that CapEx and OpEx must be lower than the revenues coming in and it should be serving a purpose for the users. Pay attention, Value and Profit make the system sustainable! Value and profit make the system growable! It should make money! That is something that most cities overlook because they think that the benefit will outweigh the revenue, but it will not. Revenue matters in the long run, and the benefits matter up front. Up front we want buy-in and residents to love it. In the long run, we need sustainability, so it does not bleed money in expenses.

Does it solve a problem for the resident? So, they see the value in it? Will they pay for it? When you start a business, you need to answer these questions. Figure out the price point. Some cities can put a tax out there to pay for the system, but that is not popular in most cities.

One example that I love in this document is the cable companies. They saw broadband up to 1

Gbps as silly, but they really didn't want to upgrade their system unless they had too. Guess what? They had to! BUT, not until they got competition in first form the likes of Verizon, AT&T, and Google running fiber all over the place for a very reasonable cost to the consumer. It paid off. Now all the cable companies are touting higher speeds. They see value because they were losing market share. That's amazing when you realize they had a monopoly for years in their neighborhoods. They had no competition, but the need for broadband and cheaper video hurt them. They thought they could control the market, but in the USE the market started swinging back with DirecTV and fiber competitors. They suddenly must be strategic.

OK, let's get back to the business of broadband. Once you build it, expect competition! If your business model works then more and more people will do it, just look the cable companies. While they had a monopoly for years, they got lazy. Now they have competition, and it's hurting them. Not just from fiber, most of the younger generation realized that they have a smartphone that can do anything the cable company can do, so why pay for both? Get the picture? The landscape has changed, and if you just look at other companies that do what you do, then you only see part of the competitive landscape.

This is happening with everything because broadband is the pipe to end all pipes. It could be through fiber or wireless, but the internet has opened doors to everything unless you live in China, then it only has a few open doors and a lot of blocked websites.

There is one thing that almost any city can agree on. You need broadband in your city to compete. The question is, "How do we get it here?" This is where you can look at other examples of successes and failures.

Google Fiber was supposed to be the knight in shining armor, but they stopped. They say they are going to wait for wireless, probably CBRS, but to be honest, there are plenty of bands and products that could deliver broadband now. I think that Google realized the profit model was not what they had hoped for, but they never said that officially, they just stopped, laid off a bunch of people, changed some leadership in the company, and they started saying wireless would save the day.

More on Google fiber stopping http://www.sacbee.com/news/nation-world/national/article110655177.html, this is a great article because they talk about all the pain points that Google saw in the real world. TV is expensive, incumbents have more control of the poles than they ever thought possible, and maybe wireless will be easier. They also say demand is not there. I don't get that excuse at all! It seems lame. I think they should say competition is fiercer than they thought. That is more like it. For me to sum it up, Google should have cherry-picked markets that didn't have too much competition. They should have focused on tier 2 cities that would not only have appreciated their presence but don't have an alternative. I live in Pennsylvania, where the cable companies rule. The rules here don't make it an ideal state to deploy much of anything, but it has plenty of underserved cities that not only want broadband, but they need it to survive. The smaller cable companies will not make the investment until they have too. They won't spend the money. This is an ideal target for someone like Google Fiber to deploy. But, alas, I am dreaming.

Where was I? Oh yes, economic development. Broadband is a foundation for economic development. We know that businesses need broadband to survive, but how do they get it? Many cities have a dig once policy, so if someone lays broadband, then many people need to get it in while he or she can. This really helps to get things moving and keep the competition behind you if they are late to the game. Fiber companies and deployment companies win! They can lay out the dark fiber and sell it later, not a bad model.

They also cover the 3 models that cities can look at.

Primary

This is where the city takes the lead by using public facilities to roll out the fiber and makes the investment to deploy. They use their assets to mount it. They are the provider. They may partner with someone, but generally, the city runs the business and takes the credit.

An example of this is the Chattanooga, TN, network. In 2010 the city decided to have a gigabit network available for homes. They rolled it out, and since then Volkswagen and Amazon both expanded to that area. We give the broadband credit since it was there, and they took a chance to deploy. Their model served other cities like Wilson, NC, and Leverett, Ma.

They also talk about Ammon, Id, who also built the gigabit backbone. They decided to provide the gigabit backbone because the local telco would not spend any money. The city was worried because they would need 50% of the market share to make it pay for itself. Guess what? They got 70% of the market share! When incumbents get lazy, there is a great opportunity!

Huntsville, Al, owns the electric utility. This provided them with the means and foundation to deploy broadband quickly and with an experienced player. They put in the backbone and leased it to Google Fiber. This is a win-win because Google didn't' have to deploy the backbone, the fiber is there and ready to use. They could move ahead quickly. The city-maintained control and could make money off it right away with a large customer waiting. The negative is that people perceive the fiber being built with city funds, but it worked! They had a utility, an income plan, a customer, so why not do it? They can lease the fiber to anyone, so they are not bound to only one customer, but anyone who wants it or needs it. Awesome! Learn more at https://www.gru.com/GRUComFiberOptics.aspx.

Santa Monica, Ca built out their network without a municipal department. They did it by connecting public facilities then expanding from there. They have a dig once policy, and when someone would dig, they would lay fiber. Learn more at http://ilsr.org/wp-content/uploads/2014/03/santa-monica-city-net-fiber-2014-2.pdf. They took the slow approach, one that would not have any upfront costs but would remain steady and efficient.

Partial

Usually a public-private partnership, PPP, that the city supports and endorses, but the private partner will be the one doing the heavy lifting and running the business. This would rely heavily on the partner, and the city would give complete support and take some of the credit, but the private business would take the profits. Partnerships matter here more than anywhere. If the partner is an ISP (Internet Service Provider) or a nonprofit, they need to be sure they can do what they say they can do. It matters big time and reputations are key.

ISPs are everywhere, not all of them look good, but many of them provide broadband to the home or local business. They often are rooted in the community if they are local and they want to succeed. However, not all of them do, I'll point that out farther down.

If you wonder about the nonprofit, I will give you an example of one that I worked on personally. In York, Pa, there is an organization called Crispus Attucks Association that sponsored an initiative to connect the local schools up to wireless broadband. While it went well, they are a shining example of a roll out to the schools in York County. This was one of the first of its kind. While it was later replaced with fiber, it's an example of how a nonprofit took the lead to deploy broadband. This was over 10 years ago. Gigabit was not thought to be cost-effective back then.

Westminster, Md, is an example of how the community knew they need to do something to attract people from the cities of DC and Baltimore out to their rural area. Beautiful and scenic, but far from major highways. They knew they needed broadband, and decided on fiber. They hooked up with Ting, https://ting.com/blog/next-ting-town-westminster-md/, who was a smaller ISP eager to roll out fiber. The city looked at the fiber as infrastructure, like a building or bridge, seeing it as a city asset and letting Ting manage the operations and customer service and sales. The city has an asset, but little risk and they are not running the day to day business, Ting is.

South Portland, Me, laid out $150,000 upfront (http://www.southportland.org/files/7514/0682/8622/06_-_ORDER_12_-_Bid_for_dark_fiber_infrastructure.pdf) to build fiber and chose GWI, https://www.gwi.net/about/ to build it. GWI will build it and run it and give 5% of the revenue back to the city.

Cleveland, Oh, decided to work with a nonprofit called OneCommunity, http://www.onecommunity.org/big-changes-onecommunity-evolves/ who is rolled out the network and is continuing to expand into other communities to increase the reach of broadband across Ohio. They are receiving support from the US Economic Development (EDA) Grant, https://www.eda.gov/grants/, continuing the work.

Facilitator
> Where the city supports the rollout, maybe offers some rules and regulations that make it easier to get started and deploy, but otherwise, it hands off. Cities can still play a part in broadband development if they have companies in their area willing to take charge and make things happen.
>
> East Lansing, Mi, has created the "Gigabit Ready" project which pulled in many groups like Michigan State University, Lansing Economic Act Partnership, various nonprofits, commercial property managers, and anyone else who would sign up. The goal was to roll out gigabit broadband, rather obvious, right? What did they do? They looked at the LEED program and thought, let's do that for gigabit access. This lead to the creation of the Gigabit Certified Building Program, http://statenews.com/index.php/article/2012/07/msu_lansing_on_track_for_high_speed_internet, to set guidelines and requirements for buildings to add gigabit broadband. This helped Spartan-Net, (taken from the Michigan State Spartans I assume), to partner with DTN Management Co so they could roll out broadband across East Lansing and beyond!

Louisville, KY, worked with Louisville Fiber to create a website that allowed people to request gigabit service across Louisville. Why? So that lawmakers could see the need for speed, and it worked! Using the addresses they gathered, they built a layout of where the heaviest concentration was showing local officials the need. Louisville gave 20-year franchise agreements to BGN Networks, SiFi, and FiberTech. It also helped Louisville to be chosen as a potential Google Fiber City, (which means very little now).

College Station, TX, took a different approach. They put out an RFP to test the market. I personally hate this because when you're on the other side, you do a lot of work that goes nowhere, but it served the city well because they got what they wanted. Suddenlink responded by promising to put in $250,000,000 into upgrading their network to make it gigabit capable, http://www.kbtx.com/home/headlines/GigaSpeed-Internet-Soon-to-be-Offered-in-BCS-276059641.html. Suddenlink got scared of having the government compete, so they got off their lazy ass and did something. College Station could motivate these guys into action! It all worked out for the residents.

In North Carolina, the NCNGN, North Carolina Next Generation Network, formed a group of universities and cities. Wake Forest, University of North Carolina, Duke, and North Carolina State got together to work with Carrboro, Cary, Winston-Salem, Chapel Hill, Durham, and Raleigh to make this happen. This is a large group and has deep resources in knowledge, data, and money. Who saw this as an opportunity? AT&T moved in and started deploying fiber. Then, not to be left behind, Frontier Communications started their deployment. Finally, RST Fiber got rolling as well. Then Google started to deploy. Now you have all the competition to make it happen and affordable.

Connecticut did something similar where 46 communities all got together to host a gigabit conference to share their vision to become the first Gigabit state, https://yaledailynews.com/blog/2015/01/16/connecticut-could-be-first-gigabit-state2/.

Not all Deployments are a Success!
In Seattle, it would have been city owned. The idea was to get gigabit rolled out across the city

to improve internet connectivity anywhere. I don't know what the agreement was between the city and Gigabit Squared, but it seems it fell apart. Maybe the company was too small or didn't understand what the deployment would need or lacked commitment. It's not clear to me what happened, but you can read more about it at https://www.bizjournals.com/seattle/blog/techflash/2014/01/seattles-fiber-deal-with-gigabit.html?page=all. The article hints that it could be more about the private company getting financing. It also hints that neither party worked on the local buy-in.

In Utah, there was a rollout y Utopia that was failing. So, Macquarie Financial took them over; this is a financial company, not an ISP or fiber company. Macquarie offered to cover the costs of the build out. However, the plan included a utility fee of $18 to $20 per household to continue. While this doesn't sound like a lot, it is more money out of pocket to continue. They had opposition called, wait for it......... Unopia! How funny is that name? UNOPIA! I must admit, I like that name, but all the same, is there an alternative? Unopia wanted to stop unnecessary fees. I get that. Why should they pay for a rollout when they don't want the service? The lesson here is that the community didn't want broadband bad enough, so it stalled. Learn more by looking at http://www.centervilleut.net/downloads/administration/ulct_-_utopia-macppp_faqs.april2014.pdf for this older deal.

Smart City Investment

After working around the smart city sector for several years, I have come to realize that most cities have no money unless there is a dire need. They all want to see demonstrations, read the reports, and understand how to deploy. That is great, but their real need is to get the money. It may be through grants, sponsors, or programs passed. Most smart city initiatives are next to impossible to get off the ground unless the financing is in place. I don't mean just for the upfront cost, but to support it for years to come.

It's not a question of wanting to do it, but the need and the money need to be aligned. That's why larger companies can be successful; they can work on getting grants and government money. Just like smart city groups, they know how to play the game.

Cities don't just look at companies, but they also look to colleges, educational grants, public-private partnerships. They all matter because they can all be a source of financing. Without money, offers don't go far, the eventually just wither away after many people spend a lot of time spinning in circles.

The thing is, the smart city play sounds good on paper, but the reality is the city wants to be a facilitator, not the primary or even partial. They know they need broadband, but if they spend tax money to get it out there, they won't get reelected. And all the politicians want to stay in office for as long as they can.

Don't get me wrong, there are exceptions, like New York City, but most of those cities already played this out. The smaller cities and communities are going to work with newcomers to give the incumbents competition. It's the best play, but the incumbents are there for a reason, they already invested heavily in those areas, and they intend to make as much money off that investment as they can. It's the smart thing for them to do, keep the competition out for as long as they can. Even if the locals hate them. The incumbent, generally a cable company, won't care until they have competition. They will get all the money they can for as long as they can, they have shareholders that want that to happen. Before you judge them harshly, let me ask, what would you do in that situation? Would you lower costs

for no good reason and cut profits? Or……
would you get as much as you can for as long as
you can. Don't judge too quickly.

Who buys broadband, really?

If we look at how broadband, gigabit especially is distributed, then what would we see? It takes a community. If one person wants it, too bad. If a community wants it and they want it bad, then it's going to happen, eventually. That is something that the cable companies missed. They were so hell-bent on selling what they had that they could have missed this opportunity. They eventually were forced to upgrade and listen. Not they see the benefits of rolling out an all internet access system. They are going to save money on tariffs to run throughout the city. They are going to start scaling back their reliance on networks. They are going to let other providers deal with paying for network access and TV show. They are not stupid. They turned sour grapes into fine wine. Give them credit. They started looking at the big picture. At least Comcast did, and the others followed suit.

Generally, the customers always start with businesses. Large businesses need broadband, but they generally work with a large ISP, like a Verizon or AT&T. That's the thing; they don't' really care about the local ISP, they work with a larger company because they can afford to and because they generally need the gigabit speeds before anyone else.

Then comes the smaller businesses, they need the local ISP. They can't afford to install the large pipes. They are the customers that are a key to growth in the communities. You see, when politicians say that small business spurs local growth, they mean it. The small businesses support the local businesses; they have the relationships in the community. Whereas the larger companies employ a lot of people, but they rely on the employees to spur local businesses. Larger companies, for the most part, care little about the local community, unless it's the owner's hometown. However, 90% the people that work at these companies care about the hometown a lot.

Then comes the residents. Here is where the local ISPs compete. Generally, most people get access from on company. So being first can be a good thing. Most people don't want to switch. That is until the local company gets too expensive or the service is so poor that people are desperate for change. Nashville TN was a good example of this when they welcomed Google Fiber with open arms. The local ISPs, Comcast and AT&T, did all they could to slow and stop Google from getting in. They were successful in some areas, and there was nothing the city could do about it. The incumbents had rights to almost all of the poles, and they were not about to let Google ruin that.

Pricing matters!

Yes, it does! Gigabit broadband went from $7,000 a month to $70 a month in a matter of a few years. Good for consumers, tough on the provider. When we get gigabit out to the masses, it will eventually become a commodity, but you still need to get it to the people. It could be fiber or wireless. People are willing to pay, but the providers will need to offer more than just access. Years ago, it was video like TV shows and on-demand movies. Now it's internet access, wired and wireless, and let the people choose what they want from there.

The thing was, we had to start, businesses needed broadband, and they got it. Then everyone else wanted it, and it is soon going to be everywhere. We need to be connected. The next question is how? Wired or wireless? While the smartphones are a part of our everyday lives, do they really need gigabit? Does our laptop or tablet need more than Wi-Fi? Ask yourself; my opinion changes too often.

When Incumbents fight Back!

Yes, the incumbents fight back. Not always in a way that makes sense. It would make more sense if they would just build out a network, but many go to court first. Many whine or say no one wants it. Most just criticize.

Look at Monticello, MN, and the city-owned Fibernet. The city did it to spark competition. This really ticked off the local Telco, TDS, which first took Fibernet to court, and lost! Then they decided to build their own network, which if they would have done that in the first place none of this would have happened. TDS had to get off its butt and move! Then Charter, another incumbent, slashed their prices to $60 a month for access. The moral of this example is that the city's plan worked perfectly, they increased competition and forced the incumbents to do something, which is what they would not do before. Before Fibernet, they did nothing. Lesson learned!

Then there was Lafayette, LA, who built the network through LUS Fiber, (Lafayette Utility Service), only to be criticized by Reason.org in a statement, http://reason.org/files/municipal_broadband_lafayette.pdf, showing that they fell short of predictions and have debt. Welcome to the world of business, it takes money and patience. However, did LUS overlook the business principles when planning? Did they get a commitment from the community to purchase it? Apparently not. LUS says they are cash positive, Reason.org says they are not. To be honest, I am not sure what the real deal is here.

Let's be honest here; Cable companies didn't care about internet access, even when the customers were begging for it. They didn't care until they had competition. They saw AT&T and Verizon offer broadband and realized that there is a market for it. Now they are becoming ISPs; I mean true ISPs are offering more and more bandwidth. Comcast rolled out Wi-Fi successfully. Now that they know there is a need and people will pay for it, they are rolling it out. That's because they are no longer a monopoly in many areas.

Google Fiber put some fear into them. So much so that they started becoming a thorn in the side of Google. In Nashville, I talked to a friend who saw the local cable company and AT&T do all that they could to block Google from attaching the fiber to their poles. Do you blame them? NO! It's the name of the game. While you may think this would not stop Google, it did. They began to see the realities of competition, petty fights, permitting, and pole acquisition. It costs money before you make a dime. A completely different model than what they're used to. So, what we see now is Google Fiber on hold hoping wireless is cost effective. They will see that site acquisition is a killer there as well. When the site acquisition costs are more than all other expenses together, you see why small cells did not roll out by the masses.

What about Wi-Fi?

Well, we all love Wi-Fi. It is a value-add, right? Does it add value?

Well, Philadelphia and Seattle had failed Wi-Fi rollouts, highly publicized and ugly. Should we let this discourage is or should we learn from these disasters? I say we learn!

- http://www.zdnet.com/article/seattle-ends-free-wi-fi/
- https://arstechnica.com/gadgets/2008/05/philadelphias-municipal-wifi-network-to-go-dark/

While I am a fan of what's coming out soon, like the CBRS and expanded Wi-Fi, several cities have successful programs. Philly just didn't have the commitment to do something like this. Luckily, they had Comcast who picked up the slack. Comcast did a great job in Philly, especially since the city had no intention of putting any money into it.

San Francisco and San Jose, Ca did a great job with their Hotspot 2.0, https://www.pcworld.com/article/2449160/free-wifi-networks-in-sf-san-jose-join-hands-through-hotspot-20.html, and Passpoint, https://www.wi-fi.org/discover-wi-fi/wi-fi-

[certified-passpoint](), programs, really the same program. This was a big win for Ruckus, but the cities new the residents needed connectivity, so they acted and provided what people can use in both cities by registering once.

Boston rolled out the "Wicked Free Wi-Fi" covering specific neighborhoods in an effort to increase downtown broadband usage. The city already has a fiber backbone, so why not extend it to the citizens via Wi-Fi?

Blacksburg Va has an amazing Wi-Fi system that was rolled out with crowdfunding by Techpad. Techpad is a local c0-working and hacking community that raised $90,000 to make this happen.

Let's not forget NYC, the city that rolled out Wi-Fi where the old telephone booths used to be. They put access points in every kiosk they put out. They give it away for free for tourists and residents to use to save on data usage on their smartphones. The LinkNYC project is the same as London's LinkUK project. They both roll out the kiosks, which are really cool, have internet access and emergency call buttons, as well as Wi-Fi hotspots. Both cities rolled out hundreds of these units throughout the cities to create an amazing Wi-Fi system and an attractive kiosk that elevate it into the elite, smart city status.

They look great! They make money through advertising and services. They are a win-win for any city.

Then there are all the cities, communities, and states that do nothing. Too many to mention. You know who you are. Yet, people continue to live in cities that have no initiative to improve. Why is that? I intend to move because, in my city, they do little, poor planning for the most part. In these areas is an opportunity for private companies to step up and try to get something rolling. Each city has different rules, so it may be too much effort and money to deploy in these cities. If that is the case, then you need to find a city that will work with you on economic development and build there. Don't' waste your time on areas that can't or won't work with you.

Some States Prohibit Public Networks!

What about the states, they certainly would not stop the city from building a network, would they? Oh yes, they would! As incredible as this sounds, it is a real thing. Some states, in fact, many states have laws in that stop city ownership or control broadband roll networks. Our friends at BroadbandNow has a website at https://broadbandnow.com/report/municipal-broadband-roadblocks-by-state/ that covers states with laws about broadband.

Colorado is a great example of control. The state that allows marijuana sales would not allow their cities to partner with Google Fiber to roll out broadband. It took an election to overturn the law.

It doesn't always work out like that. In North Carolina and Tennessee, the FCC tried to have those state broadband regulations overturned, but the FCC lost. The laws remained.

The states with these laws are Alabama, Arkansas, California, Colorado, Florida, Louisiana, Michigan, Minnesota, Missouri, Nebraska, North Carolina, Pennsylvania, South Carolina, Tennessee, Texas, Utah, Virginia, Washington, and Wisconsin. The laws are geared towards cities that want to own or partially own broadband networks. I think the idea is that broadband should be competitive.

This parallels what many states are doing with small cell deployments for the carriers. They have been passing state laws that allow the carriers to roll out their small cells without the local municipality slowing them down. The CTIA has done a great job lobbying the states to make life easier for the carriers at the expense of the local cities.

City Strategies for a Broadband Initiative

There are several strategies for any city to start the initiative.

1) Do you want to undertake this effort? Is this something that you need for the city? Can you build a story around it?
    a. Then do the research, get the buy-in, and lay out the options.
2) Weigh out your options to get started? This is the first step.
    a. Will you do it yourself or gather partners?
    b. If you partner, which you probably will, who do you reach out to? Think about whom it will benefit. It could help the universities and the local businesses. Start with them first because they have the most to gain. Broadband attracts business and talent, both of which help the local universities and cities.
    c. Look at what's been done. Are there groups, like Next Century Cities and Smart Cities Council,

that you can reach out to for help? Are there other cities that would offer advice? Of course! Do some research and look them up.
3) Once you have a partner and maybe a high-level plan, then what?
   a. Will you own the network?
   b. Will you want local businesses to build out the network?
   c. Will you offer incentives, like easing the permitting process, offering city assets, or other things to promote local business to deploy?

What will the future hold?

One thing that is changing the way we look at broadband is SDN, Software Defined Networking. As AI improves, and by AI, I mean augmented Intelligence, it could mean Artificial Intelligence. What is the difference?

Augmented Intelligence is where we enhance the decision making of the machines; they only make corrective or guided decisions. We are there today. Artificial Intelligence is where the machines make a choice altogether, without human intervention and they learn on their own.

How does this define networking? If a router or server or controller can decide what the packet is used for and where it should go, it can choose the fastest route with the growth of IOT, Internet of Things, the networks could get overwhelmed with small packets. We have been optimizing the networks for larger packets for video and large data transfers. Now we are going to look for a matching to send out kilobytes instead of gigabytes, and it's going to create havoc for these optimized networks. What if you had virtual networks optimized for a specific purpose, like video would go one way and route through a specific system and IOT data could route another way that is very low latency? Sound familiar? Of course, it's 5G standards that want us to route the traffic this way, and we need to do all that we can to make it happen. The technology has arrived, ow we need to deploy.

However, I am not telling you this for the wired networks, that have all the cool features waiting to be deployed on a mass scale. I'm telling you because the wireless networks must get there.

What? Aren't wireless networks there?

Not really, you see most wireless backhaul systems come from one source at a time. This must change. The wireless connection for broadband has to come from more than one source. How? We have UE backhaul; soon this will connect almost everything for broadband. It should be able to simultaneously connect to multiple sites simultaneously and use all of them for backhaul simultaneously. This is how we break the barrier of limited spectrum.

That is already here; it's called carrier aggregation. It is used to connect to the end user using multiple bands of spectrum. Here are some examples:

- LWA – LTE and Wi-Fi Aggregation, this is where the LTE licensed spectrum works with the Wi-Fi spectrum to have one big

pipe. While this sounds easy and cool, Wi-Fi and LTE do not play well together for so many reasons. It has not worked out as well as the carriers had hoped. Lots of technical issues with the Wi-Fi reliability.
- LAA – Licensed Assist Aggregation is where LTE licensed would work with LTE-U, (LTE in unlicensed band the same as Wi-Fi). The 2 LTE carriers would work well together, and I would be one big pipe. However, unlicensed is still sharing spectrum, and it may run into similar problems as Wi-Fi. The big difference is that LTE loves LTE and they work well together when being aggregated together.

There is standard aggregation that the carriers are already using to get all the LTE bands to work together. We don't hear about this because it is in use today and it's the secret to making the LTE connection look like 100Mbps. The carriers are getting all their existing carriers to work together to look like one large data pipe directly to your smartphone, so you can see your favorite cat videos on YouTube. You can learn more at https://wade4wireless.com/2015/09/08/lwa-laa-lte-u-and-wi-fi/ if you want to learn more about the differences.

Acronyms and Definitions

- **AI** - Artificial Intelligence or Augments Intelligence.
- **CBRS** – Citizens Broadband Radio Service – in the USA this is 3550MHz to 3700MHz, often referred to as the 2.5GHz spectrum. Learn more at https://www.leverege.com/blogpost/what-is-cbrs-lte-3-5-ghz
    a. **ASA** - Authorized Shared Access
    b. **PAL** - Priority Access Licensed
    c. **LSA** - Licensed Shared Access
    d. **GAA** - General Access User
- **CCI** – Crown Castle Incorporated
- **CLEC** - Competitive Local Exchange Carrier
- **Cmwave** – spectrum in the 3 to 30 GHz range and will most likely be used in 5G for the fixed spectrum but could have mobility potential.
- **CRAN** – Centralized RAN
- **cRAN or C-RAN** – Cloud RAN
- **DAS** – Distributed Antenna Systems
- **FTTH** – Fiber to the Home
- **FTTSC** – Fiber to the Small Cell
- **FTTx** – Fiber to the Anything
- **FWA** – Fixed Wireless Access
- **IOT** – Internet of Things

- **KPI** – Key Performance Indicators
- **LoRaWAN** – Long Range Low Power WAN
- **LOS** – Line of Site
- **LPN** – Low-Power Network
- **LPWAN** – Low Power Wide Area Network
- **LTE** – Long Term Evolution
- **LTE- U** – LTE Unlicensed, generally license-free spectrum in the 2.4GHz and the 5.8GHz ISM bands
- **MEC** – Mobile Edge Computing
- **Mmwave** – spectrum in the 30 to 100GHz range that will be part of 5G and most likely used for fixed wireless.
- **MIMO** – Multiple I Multiple out (antennas)
- **NB-IOT** – Narrowband Internet of Things
- **NLOS** – Near or No Line of Site
- **PaaS** – Platform as a Service
- **PoE** – Power over Ethernet
- **POTS** - Plain Old Telephone Service
- **PTP** – Point to Point
- **PTMP** – Point to MultiPoint
- **PTT** – Push to Talk
- **RAN** – Radio Access Network
- **ROW** - Right of Way
- **SaaS** – Software as a Service

- **SAS** – Small Cell Antenna Systems, like DAS systems but with small cells only.
- **SCaaS** – Small Cell as a Service
- **SDN** – Software Defined Networking
- **SISO** – Single in Single out (antennas)
- **UE** – User Equipment, like a smartphone
- **Wi-Fi** – Wireless Fidelity, generally license-free spectrum in the 2.4GHz and the 5.8GHz ISM bands
- **WiMAX** – Worldwide Interoperability for Microwave Access, based on 802.16 set of standards, learn more at https://en.wikipedia.org/wiki/WiMAX

Thank you!

Thank you for your support. I pray that it serves you well.

I want to thank you for your thirsting for technical know-how. The people that want to learn more will grow in every way. I am here to serve that need.

If you need one on one consulting or specific reports, feel free to reach out at wade@techfecta.com or wade4wireless@gmail.com for direct support.

Copyright

First Edition © 2018 by Wade Sarver. All rights reserved. No part of this publication may be reproduced, stored in a retrieval system, or transmitted in any form or by any means, electronic, mechanical, photocopying, recording, scanning, or otherwise, except as permitted under Sections 107 or 108 of the 1976 United States Copyright Act, without the prior written permission of the author.

I am not a lawyer or an actively certified safety expert. This book is completed based on research and my experiences. Safety processes and procedures are constantly updated and improved over time. The material contained is for reference only and may include products, information, or services by third parties. I do not assume responsibility for any third-party material referenced in this book.

This document is a guide to help people and not a guarantee that you will do everything properly. By reading this, you agree that myself and my company is not responsible for the success or failure of your business decisions relating to the information presented in this guide.

www.wade4wireless.com

www.techfecta.com

Cover and design by Wade Sarver

About Wade

Chief Technology Analyst for TechFecta.

Solutions Consultant service tech investors and tech businesses.

Tech author, blogger, podcaster.

TechFecta, Tech consulting for the real world.
www.techfecta.com

Blog and podcast available at
www.wade4wireless.com **if you want to follow.**

Link up with me on LinkedIn, https://www.linkedin.com/in/wadesarver/.

Reach out to Wade on LinkedIn or at Wade@techfecta.com or wade4wireless@gmail.com to send feedback.

Twitter @wade4wireless, https://twitter.com/Wade4Wireless.

Learn more!

The blog at www.Wade4Wireless.com to offer real-world knowledge to wireless workers and investors.

TechFecta is here to help workers, businesses, and investors gain clarity on the direction technology is heading from where it is today.

Other products include:

- The New T-Mobile Consolidation Report https://wade4wireless.com/2018/05/13/the-new-t-mobile-consolidation-report/
- The 5G Deployment Plan Book https://wade4wireless.com/2017/01/30/the-5g-deployment-plan-book-release/
- The Wireless Deployment Handbook for LTE Small Cells, CRAN, and DAS.

https://wade4wireless.com/2015/11/12/wireless-deployment-handbook-for-lte-small-cells-and-das/
- **Learning 5G in the Real World.** https://wade4wireless.com/2017/09/18/learning-5g-in-the-real-world/

www.ingramcontent.com/pod-product-compliance
Lightning Source LLC
Chambersburg PA
CBHW071519220526
45472CB00003B/1080